THE REALM OF THE WILD

LIFE AND DEATH ON THE AFRICAN SAVANNAH

SMITHMARK

THE REALM OF THE WILD

LIFE AND DEATH ON THE AFRICAN SAVANNAH

Texts and photographs
Michel and Christine
Denis-Huot

Editor
Valeria Manferto De Fabianis

Graphic design
Anna Galliani

Translation
Cathy Muscat

Contents

This edition published in 1997 by Smithmark Publishers, a division of U.S. Media Holdings, Inc., 16 East 32nd Street, New York, NY 10016.

SMITHMARK books are available for bulk purchase, for sales promotion and premium use. For details write or call the manager of special sales, SMITHMARK Publishers, 16 East 32nd Street, New York, NY 10016; (212) 532-6600.

Produced by: White Star S.r.l.
Via Candido Sassone, 22/24
13100 Vercelli, Italy.

ISBN: 0-7651-9330-2

Printed in Italy

1 The highest density of lions per square mile can be found in the Masai Mara reserve in Kenya between the months of June and October. In addition to the resident prides are the nomadic lions from the neighbouring Serengeti reserve that follow the great migration of herbivores to this area. These felines, however, have almost disappeared from the Amboseli reserve, having been exterminated by the Masai.

2-3 Jousting between elephants of the savannah is generally friendly. Both protagonists begin by touching each others' trunks, then coiling them up. Then they stand face to face, tusks interlocked and push back and forth, tussling. The level of noise is impressive. Sometimes, violent fights occur over a female. These more serious confrontations can last longer than eight hours.

4-5 These male dwarf flamingos are taking part in a mating parade. A handful of them had gathered around a receptive female and were subsequently joined by hundreds, if not thousands of other flamingos; they crowd closer and closer together, their heads erect, all moving in the same direction. These strutting males whose blood flows just beneath their skin, form a sea of red among the other birds.

6-7 The Nile crocodile is one of the biggest living reptiles. They once grew to lengths of 8 metres, but now, crocodiles measuring over 4.5 metres have become rare. They can be found on the banks of the Mara river in Kenya and the Grumeti river in Tanzania.

8-9 The future of the cheetah is in jeopardy. In Africa, there are between 5,000 and 15,000 cheetahs, one thousand of which are in Kenya. This number continues to dwindle and trade in their skins has been one of the key factors in their decline. Reduced habitat and persecution by farmers and cattle-growers continue to threaten its existence. Moreover, its resistance in the reserves has been weakened because of the increased competition with other predators such as lions, and the increasing disturbance caused by tourists. Solutions for safeguarding this beautiful feline are not simple.

10-11 The cheetah spends so much energy hunting that it must rest at least 15 minutes before eating, even longer if it has been hunting during the hotter part of the day. It eats quickly, always on the alert as it cannot defend its property against other predators such as hyenas, baboons, lions and jackals.

INTRODUCTION

East Africa, home to the richest savannah regions in the world, is also distinguished by the incredible diversity of its landscapes: glacier-capped volcanoes, salt lakes and hot springs, stretches of equatorial forest rising up mountainsides, mangrove trees, white beaches of fine coral sand, windswept, fertile high plateaux richly cultivated by the people who live upon them, and torrid, low-lying plateaux that are virtually uninhabited. This extraordinary geographical diversity is linked to the Rift Valley, a gigantic canyon stretching from Lebanon to Mozambique, produced by the awesome geological activity that began in this region thirty million years ago.

The high plateaux in the east were formed when violent subterranean forces caused hot matter of low density to rise beneath the continental crust, causing it to bulge. These plateaux then fractured, magma bubbled up to the surface, the huge blocks of crust plates buckled and the centre collapsed, forming the Rift Valley.

This zone is still geologically active and is currently drawing further and further apart. It is quite possible that the canyon will develop into an ocean over the next few million years.

These high plateaux and volcanoes have formed a kind of barrier over which clouds cannot pass. This has caused the area east of the Rift Valley to gradually become arid, while in the west a hot and rainy equatorial climate persists. The vegetation in the east has thus been altered. In the high altitude areas the humidity has allowed for the development of a grassy savannah, a kind of prairie dotted with trees such as the umbrella acacia. Numerous wild herbivores and large carnivores including lions, cheetahs, and hyenas feed here. In the hot, dry climate at lower latitudes, the landscape predominantly consists of prickly brush and bushy steppes with thickets of thorny dwarf

acacias out of which tall trees, generally baobabs and giant acacias, emerge at wide intervals. Rising out of the midst of this vegetation are the ochre, brown and red columns of giant ant-hills. Unlike the savannah, this biotope is very hostile to man and is known as "nyika", which means wild and deserted. Birds thrive here in abundance, but not many mammals can withstand the conditions, apart from impalas, gazelle-giraffes, dik-diks, giraffes, black rhinoceroses and leopards. These two forms of vegetation - savannah and semi-desert steppe - are also home to tribes of herdsmen such as the Masai, Samburu, and Turkana. The tropical climate of the savannah region is clearly divided into dry and rainy seasons. The morphology of the trees and bushes here is characteristic of such a habitat. The roots of the acacia trees penetrate deep into the earth, spreading radially to draw out the maximum amount of moisture. The thick, impermeable tree bark prevents evaporation. Unlike acacias and the various species of thorny plants which can survive on little water, baobabs and euphorbia are able to store enormous quantities of water in their trunks and branches.

During the course of their evolution over the millennia the trees have grown taller keeping a large portion of their foliage out of the reach of foragers, forming, for example, the umbrella-shaped canopy typical of certain acacias. Thorns have also appeared for protection from leaf-eaters, though they are powerless against the tough tongues of giraffes and black rhinoceroses. To improve their protection against animals, some acacias have developed other systems of defence. The whistling acacia, for example has developed outgrowths at the base of its thorns. These hollow spheres are colonised by ants that feed off the tree's secretions.

13 Elephants regularly come to the salt-marshes to eat the soil, even though they have trouble digesting it. The soil gives them stomach pains and they even run the risk of poisoning themselves because of the excessive amounts of potassium in the plants which they ingest along with it. But by eating quantities of dirt, they gain the iron, magnesium and calcium necessary for the growth of their tusks.

14-15 The African buffalo is a rather unpredictable animal and can be very dangerous when it charges. Solitary individuals are even more aggressive as they tend to be old male buffaloes that have been chased from the herd. Their impressive horns, highly prized by hunters, are deadly weapons against predators.

16-17 The male lion exudes dignity and instils fear through the noble bearing of its magnificent head, its regal mane and its deep, rumbling roar. A lion's mane begins to grow when he is two years old; its colour varies according to the geographical region and ranges from pale yellow to deep russet brown.

When an animal touches the tree hundreds of ants come out of hiding and rush toward the aggressor to repulse it. Other acacias let off volatile substances when nibbled on by antelopes, which trigger the release of tannins by other acacias farther away. These tannins render the trees inedible and thus protect them from herbivores' teeth. Animals, too, have adapted themselves to their environment. A giraffe's elongated neck gives it a height advantage of approximately 5 metres (16 feet) over its competitors. Antelopes have had to adapt to hotter temperatures and the absence of water. Those who live in desert regions, like the gazelle or the oryx, cool themselves by increasing the frequency of ventilation cycles rather than through perspiration. In fact, animals which cool down by perspiring, like the buffalo, cannot limit their water loss, even when water is scarce; their habitat is thus limited to humid savannahs or to areas in which water is available. Gazelles only pant when their body temperature rises about 43°C (118°F) while most other animals cannot survive beyond 40°C (112°F). However, even in the case of gazelles, the temperature of the brain must not exceed 41-42°C (110°-115°F) otherwise irremediable lesions will occur. Warm blood arriving to the brain must be cooled down. External carotids at the base of the cranium are divided into hundreds of minute arterioles which join up again after passing through a veined sinus which cools the blood between 3 to 6°C (8° to 16°F) through water evaporation in the nasal mucous. The temperature of the blood drops even further when the animal pants quickly. Zebras provide another example of adaptation. All wild equids are descendants of the same North American ancestors, but only African equids which first appeared on the savannah have stripes.
Though each zebra has its own distinctive

stripes, the role these markings play in identifying individuals cannot alone account for their appearance. The zebra is the only large mammal immune to the sleeping sickness disease that is carried by tsetse flies, to which its cousin the horse is extremely vulnerable. Experiments have shown that the tsetse, which is vision-directed, is not attracted to alternating bands of black and white. This explains the mystery of where the zebra got his stripes - or why at least. And if other mammals of the savannah are not striped it is because they have developed immunities to the tsetse, after having lived with these insects for millions of years. Because equids arrived on the African continent much later they had not had the time to evolve and adapt to their environment; their mortality rate was around 90%. To ensure their survival nature offered a solution in the form of striped camouflage.

Today, the great wild fauna of Africa has all but disappeared outside of the national parks and reserves. When the first Europeans arrived, they quickly discovered that the savannahs were home to an infinite number of animals. In the XIXth century Arabs trading in slaves, elephant tusks, rhinoceros horns and animal skins decimated wild games stocks on the continent to the point where, by the end of the century, many species were on the verge of extinction. The majority of the region then passed under German or English protection. Having learned from the scandalous elimination of game by South African and North American colonists, German and English officials took steps to protect animal life. Largely insufficient, these steps could not completely prevent poachers and hunters from continuing their destructive onslaught. The great bovine epidemic of 1890 paradoxically played a role in protecting wild fauna.

18 top Leopards hunt at nightfall. These sleek wild cats demonstrate an uncanny ability to adapt to their environment. They are found in tropical rain forests, searing deserts, low-lying marshlands and at altitudes of more than 4000 metres.

18-19 Each leopard cub lays claim to one of his mother's teats at birth, which it impregnates with its own scent in order to be able to locate it quickly and thus avoid fighting with its siblings or scratching its mother.

20-21 Hippos spend most of their time in the water, but, curiously, they eat very few of the aquatic plants which grow abundantly around them. Instead, they wait for nightfall when they emerge from the water and graze on the grasses that grow by the shore.

22-23 During the dry season elephants spend much of their time in the swamps. When they are sick, injured or old, they seek refuge here because of the water and the tender, easy-to-chew vegetation. When they die their bones sink into the water or are hidden among the vegetation. All it takes is a dry spell to uncover a concentration of their skeletons, a phenomenon which no doubt gave rise to the myth of elephant burial grounds.

23 top In a few hours several hundred vultures will converge on this carcass. Their talons are not as powerful as those of eagles and they consume their food where they find it, thus clearing the region of cadavers and preventing the spread of disease.

Domestic herds were hit hard by this disease and many herders, deprived of cattle, abandoned their pastures, creating vast "no man's lands" that were quickly overrun by tsetse flies. Wild fauna was also affected: 95% of gnus and buffalo were decimated in the space of two years, while lions were forced to become man-eaters. Wild animals rapidly took possession of the tsetse zones, creating what were, in fact, huge game reserves. In 1933 the decision was made to create national parks and reserves, but these were not properly established in Kenya and Tanzania until later, between 1948 and 1968. The largest area of protected ecosystem in the region stretches over approximately 30,000 square kilometres. It is not confined by natural boundaries and is formed by the Masai-Mara reserve in Kenia, Serengeti Park and the Ngorongoro Crater conservation zone in Tanzania. It is here that the world's only remaining mass migration takes place every year, involving about two million herbivores, 1.3 million of which are gnus. These vast numbers are the result of the eradication of the bovine disease through veterinary treatments, which in turn wiped out the virus among wild animals in the sixties. Heavy rains followed the great drought of 1971, which had stemmed gnu and buffalo population growth, as the size of an animal population is directly linked to the richness of the soil that must support it. The ecological balance is always fragile in protected zones. Intensive poaching has seriously jeopardised some species: the rhinoceros, for example, is a prime victim hunted for its horns, which, when ground into powder, fetch astronomical sums on the international market for their supposed aphrodisiac powers. Rhinoceros horns are also used in Yemen to make dagger handles. Today there are fewer than 4,000 black rhinos (400 of which are in Kenya) and only a handful of white rhinos, imported from South

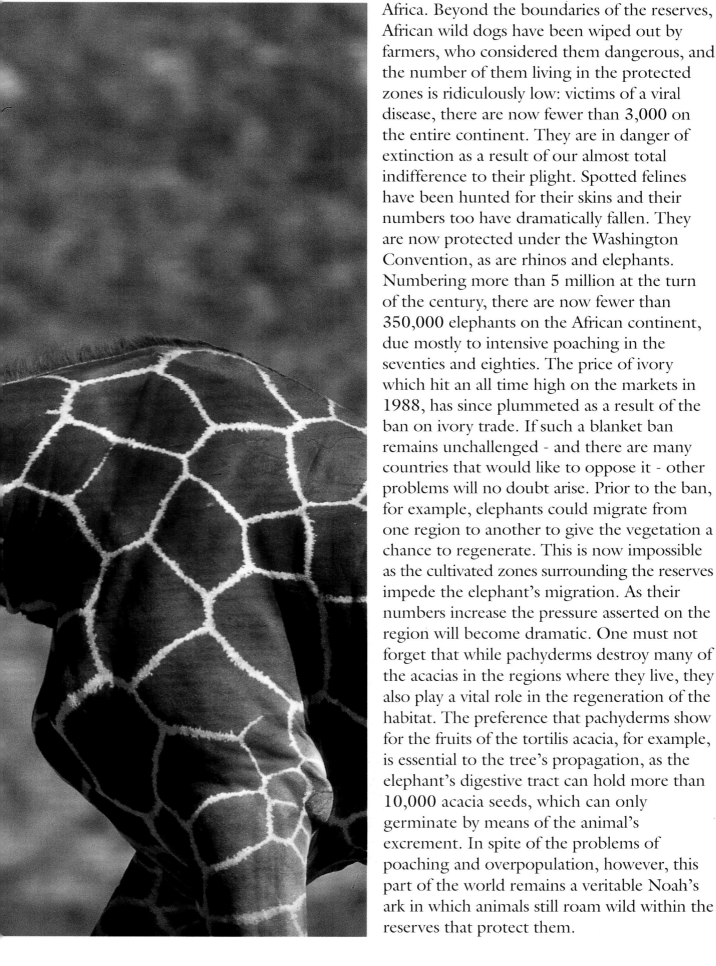

Africa. Beyond the boundaries of the reserves, African wild dogs have been wiped out by farmers, who considered them dangerous, and the number of them living in the protected zones is ridiculously low: victims of a viral disease, there are now fewer than 3,000 on the entire continent. They are in danger of extinction as a result of our almost total indifference to their plight. Spotted felines have been hunted for their skins and their numbers too have dramatically fallen. They are now protected under the Washington Convention, as are rhinos and elephants. Numbering more than 5 million at the turn of the century, there are now fewer than 350,000 elephants on the African continent, due mostly to intensive poaching in the seventies and eighties. The price of ivory which hit an all time high on the markets in 1988, has since plummeted as a result of the ban on ivory trade. If such a blanket ban remains unchallenged - and there are many countries that would like to oppose it - other problems will no doubt arise. Prior to the ban, for example, elephants could migrate from one region to another to give the vegetation a chance to regenerate. This is now impossible as the cultivated zones surrounding the reserves impede the elephant's migration. As their numbers increase the pressure asserted on the region will become dramatic. One must not forget that while pachyderms destroy many of the acacias in the regions where they live, they also play a vital role in the regeneration of the habitat. The preference that pachyderms show for the fruits of the tortilis acacia, for example, is essential to the tree's propagation, as the elephant's digestive tract can hold more than 10,000 acacia seeds, which can only germinate by means of the animal's excrement. In spite of the problems of poaching and overpopulation, however, this part of the world remains a veritable Noah's ark in which animals still roam wild within the reserves that protect them.

24-25 Reticulated giraffes inhabit the semi-arid zones in the north of Kenya. The males cohabit in overlapping domains, regardless of the sub-species they belong to. When two of them meet, their status is determined by jousting during which they intertwine their necks and push each other with often violent blows of the head. Deadly blows with the hooves, however, are only used against predators. Combats between giraffes are graceful and spectacular and rarely dangerous, except when one of the combatants loses his footing and falls to the ground.

26-27 Lions seek refuge during the day in the shade of trees and bushes, as they fear the heat. If they have not finished with their prey before the real heat of the day sets in, they drag it with them into the shade.

28-29 Grant gazelles graze near umbrella acacias. Even one such tree in the parched desert is a blessing for animal life. Antelopes eat the leaves from the bottom of the tree while elephants and giraffes feed from the top, and dozens of weaver-birds build their nests in its branches.

WILD PORTRAITS

In the Sambura Reserve, a herd of elephants makes its way towards the river in the early hours of the morning, grazing as they go. A pair of tiny, graceful antelopes - dik-diks - appear from behind a bush. Next to them a male impala with superb lyre-shaped horns watches over his harem of a hundred females and their young. Other males graze nearby, from time to time attempting to approach, only to be run off by the "proprietor". As they move towards the river the pachyderms pass under an acacia tree, where a leopard is resting peacefully, indifferent to what is going on around it. On the other side of the water two male giraffes battle neck to neck, and head to head in an exquisite ballet. A dozen other giraffes graze a few feet away, looking up once in a while at the two combatants. The cries of hippos ring out from the river. These are the ordinary scenes of reserve life, reflecting the great diversity of species that thrive here.

The different animal societies are organised in a variety of ways: lions, wild dogs, hyenas, baboons and elephants live in groups and strong ties are developed between individuals; jackals and dik-diks stay in couples; and impalas, gazelles, topis and gnus stick together in herds. Within each social grouping, males and females of the same species behave according to different rules. Fighting, for example, is in large part a male trait. What is at risk in combat is significant: the conquest or protection of territory, defense of one's place in a hierarchy, conquest of females to mate with, or the appropriation of prey from a competitor. An animal never fights another animal simply because he does not like him! Fights are rarely to the death, except among hippos. The death of an adversary of the same species, when it does happen, is neither intentional nor particularly sought after, but only the result of injuries received. Very often the protagonists will do everything they can to avoid a full-scale fight.

In very hierarchical societies, for example, postures of intimidation and submission allow ranks to be established or confirmed and let the weaker party withdraw without having to fight.

This is very common among elephants, for example. An elephant herd is principally composed of females and their young, while male adults are excluded from the matriarchy. Day after day, season after season, herds migrate along the same trails and all the members busy themselves with the same occupations. They may stray a little from each other while eating but they take care to maintain contact by trumpeting or making high-pitched sounds that are inaudible to human beings. Elephants communicate effectively by infra sound, which can carry over great distances; this is especially true in the vast open spaces of the savannah. The oldest female of the group, the matriarch, leads. The others are all her relatives: sisters, nieces, cousins and daughters. She is both the group's guide and its memory; her companions obey her dutifully. Her knowledge of the clan's domain and the trails her ancestors followed from time immemorial is essential; she knows best where to find food and water each season and how to avoid dangerous places. Thanks to her prodigious memory her young relatives can gradually acquire this knowledge. They learn by imitating her example.

Contrary to primates, who only have social relations with members of their own clan, the females of an elephant family have connections with other groups, whose members are also close or distant relatives. During the rainy season, when food is abundant, these "families" regroup in huge herds that encourage social exchange. As soon as food becomes scarce again, however, they separate.

31 Elephants - and their tusks - continue to grow throughout their lives. A male elephant weighs an average of 5 tons and his tusks weigh between 30 to 40 kilos (65-90 pounds), compared to female tusks which weigh around 3 tons and 10 kilos (22 pounds). The largest tusks ever seen weighed 105 kilos (230 pounds) and were 3.5 metres (11.5 feet) long. These incisive hypertrophies perform many functions. They are used to peel bark, drill wells and dig up roots, and they are also used as weapons in combat.

32 This elephant is worried and lifts his trunk to sniff at a suspicious scent. An elephant's trunk, a unique organ in the animal kingdom that can reach a length of over 2 metres, is in fact an enormous muscular mass that terminates in two tactile and prehensile "fingers". Its uses are multiple: nose, hand, vacuum, sonorous alarm system, visual signal and sensory organ.

33 An elephant can use its strong trunk to tear up hardy aquatic plants, branches and shrubs, or, conversely, delicately pluck a leaf or a blade of grass.

The life of male elephants is completely different. They are forced to leave the herd at puberty. As they are inexperienced they stay close to the females, in small groups of males of a similar age. When they are between 20 and 25 years old they go their own separate ways, moving with great regularity in well-defined domains. Sometimes they are accompanied by one or two younger companions, like knights with their pages. They rejoin the females during the rutting season. After the many jousts that they have taken part in over the years, the males develop a sense of their respective strengths and a pecking order that covers food, water and mating is established without the need of any fighting.

If the structure of the population is sound, if poaching has not upset the age pyramid, young males will only briefly have the right to mate, at the beginning and end of the females' heat cycles. Males older than thirty demonstrate at certain periods of the year astonishing physical manifestations: their temporal glands become inflamed and secrete a viscous liquid with a very strong odour; the inside of their back legs become soaking wet and their penises become a greenish colour. During this rutting state - the musth - which lasts between two and three months of the year, their behaviour changes: they become very aggressive and the hierarchy is thrown into flux. These musth males take control no matter what their rank is and become responsible for the greater part of the herd's reproduction.

34-35 The female members of an elephant herd spend their lives together and the ties between them are very strong. They never quarrel and show each other the greatest tenderness. When a family encounters a group of "friends" the females greet each other by intertwining their trunks and banging their tusks or slapping their ears together, trumpeting noisily all the while. The care they show for weaker members, be they young, sick or injured, is considerable.

36-37, 36 top and 37 top
When male zebras switch
from play-fighting to real
fighting, they bite each
others' throats, necks and
legs. Their hooves, which
are composed of horny
bone and grow
throughout their lives,
are wielded like weapons,
to be feared not just for
the powerful blows they
can deliver but also for
their ability to slice into
an adversary with their

sharp edges. During these
combats the stallions
measure their blows so
as not to cripple their
adversaries. Contrary
to that which is the case
with the majority of
hoofed animals, where
the winner is the one
who demonstrates the
greatest physical strength,
male equids recognise as
dominant those stallions
who are the swiftest and
most agile fighters.

Zebras also like to fight, but they fight
to conquer and defend territory.
Their "duels", part of a complex ritual,
start during adolescence and continue into
adulthood. Stallions only stop fighting when
they start a family, which usually occurs when
they are four or five years old.
Before reaching this age they confine
themselves to groups consisting of single
young males and stallions too old to
maintain charge of a harem. When the time
comes to start a family they first choose one
and then several mares.
The colts and foals born will complete
the group. Social ties are reinforced by
the occupations practiced in common and
in particular by the act of reciprocated
cleaning, where one zebra gently grabs
the skin of the neck and withers of another
and delicately scratches it with his incisor
teeth. Each family head salutes the other
stallions living in the area by performing
a ritual of flaring his nostrils, rubbing
his head against the flank and stomach of his
comrade, and then leaping in the air to say
good-bye. When the herd is on the move
it is the highest ranking mare that leads
the way. The others follow according to rank
with the young zebras mixed in among
them. The male stays at the back of the herd
or a little way to the side. The colts and foals
leave the group when they reach maturity.

37 bottom right
Are zebras black with
white stripes or white
with black stripes? Most
scientists lean towards the
second hypothesis.

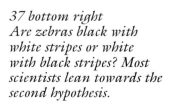

For Masai warriors the lion is the symbol of nobility and courage. It is the only wild cat that lives socially: a pride is composed of a coalition of one to six males and four to fifteen females together with their young. Contrary to other very hierarchical societies lions and lionesses have more or less equal rights. It is only during feeding time that the predominance of the males asserts itself, as they are more powerful than their companions. The females are all related to one another and usually spend their whole lives with their birth pride. Young males, by contrast, are chased away when they are around three years old. They live alone or in groups of two or three and become nomads, but stay close to the prides, checking for signs of weakness or age among the males who belong to them. The challenges launched by these youths against the resident males result in combats that are sometimes lethal. Taking power is always only temporary, however: it is estimated that the rotation of males in a pride occurs every two or three years. The roles of lions and lionesses are very distinct. Males defend the territory and the pride against outsiders. To mark their domain they urinate against shrubs and grassy patches, creating olfactory boundaries. If an intruder is caught inside the domain its holders will usually just chase him off with a minimum of aggression, but they can just as easily be merciless. In exchange for this protection the females raise the cubs, hunt and kill. There is no mating season. The first male to encounter a female on heat will stay by her side and keep the other males away by his presence. For this reason there is little rivalry between males. Mating only lasts a few minutes but it is repeated frequently, as many as fifty times over a 24 hour period. This is continued for as long as the female is sexually receptive, usually five days. During this period the couple does not hunt and barely eats. In spite of the frequency of copulation,

however, the chances of procreation are relatively slight. Next to the stealthy felines, the rhino - resembles the long-extinct triceratops - is an ungainly creature. But despite their prehistoric appearance, these herbivores, like lions, are sexually powerful: mating can last up to an hour with several ejaculations and their phalluses can reach up to 1 metre (more than 3 feet) in length. It is because of its incredible sexual power, however, that the rhino is such a lucrative victim for poachers. Rhinos lead solitary lives, use the same trails everyday and follow very precise schedules. They are fairly tolerant of their neighbours and commonly share watering-places. Outsiders, however, are systematically chased away.

38-39 Lions use a variety of sounds and facial expressions to communicate, thus minimising the need to resort to violence. When a lion roars, standing, with his head tilted down towards the ground, his flanks pulled in and his chest swollen, he is announcing that the territory is his. The ground around him vibrates!

40 top The colour of a lion's iris varies from gold to brown depending on his age. He relies much more on his vision and his hearing than on his sense of smell.

40-41 Lions often exhibit tenderness. When mating, however, the male bites his partner's neck and the female moans heavily and grimaces as if she were threatening an adversary. After copulation, the male beats a hasty retreat as the female will not hesitate to attack him.

41 Females of a pride are very attached to their place of birth, though they sometimes have to leave it when it becomes overcrowded.

This attachment stems from the fact that this was where, during their youth, they acquired a crucial understanding of their hunting grounds.

While the proud lion is the most admired inhabitant of the savannah, other animals, by contrast, have a bad reputation, the obvious example being the spotted hyena. Unattractive, with its short, dirty coat, receding hindquarters and noxious odour, it is renowned for its hideous, blood-curdling laugh, a sound which penetrates the African night. Hyenas have a highly developed social life. They live in clans of fifty individuals and are in constant communication with one another, through all manner of sounds, signs, postures and odours. Fighting is rare in the group as the hierarchy is rarely questioned: adults dominate the young and females dominate males. In this highly ordered matriarchal society, led by one or two females, the notion of couple does not exist. When a female is on heat she can attract as many as fifteen males who will fight to mate with her. Once the winners are decided they must then be accepted by the female whom they approach with the greatest caution, while demonstrating signs of submission.

Contrary to the cohesiveness found in lion and elephant groups, giraffes live primarily as individuals, even though these herbivores are often seen together in small and sometimes large groups during the rainy season. They are constantly on the move and the composition of the group changes from one day to the next or one week to the next for no apparent reason. Giraffes keep constant track of one another through their remarkable visual acuity and the extraordinarily lofty position of their heads. Males are not territorial and live together in the domains where they graze. All social relations are nevertheless governed by a strict hierarchy: young giraffes and young adult males establish their place in the group through ritualised combat.

42 top At any given time a giraffe will provide shelter and sustenance to a large number of bullock-driver birds.

42 bottom A giraffe is characterised by its long neck but, like other mammals, it only has seven vertebrae, each one more than 40 centimetres long. When it moves, it uses its neck to keep balance by displacing the centre of gravity forward or backward.

43 Male giraffes can reach a height of 5.8 metres. Their markings, which differ from one giraffe to the next, stay the same throughout their lives. Their colouring however, will grow darker as they age.

44-45 The wider a hippo can open its mouth the more it will occupy a central place in the group; only dominant males can open their jaws 150°. This enables them to show off their impressive canines, which can measure up to 60 centimetres and weigh as much as 3 kilos.

45 top These hippos have just started to fight: their mouths gaping, they exchange violent lateral blows of the head and try to dig their canines into each others' necks and stomachs. Fortunately, their vital organs are well protected by highly-resistant skin and a thick layer of fat. Adult males are covered with scars.

45 bottom
The tumultuous confrontations, accompanied by water-spitting, groaning, bellowing and heavy breathing, will only continue and grow serious if the males are of the same age. Hippos will reinforce their aggressive actions with resonant bellowing.

46-47 *In regions where the hippo population is particularly high, the dominant males must often resort to living in small mud pools, where they become much easier victims for predators.*

The hippopotamus may seem a placid enough creature, but appearances can be deceptive. This mastodon can be very dangerous both on water and on land where it can reach speeds of up to 45 kilometres per hour. The communal structure of this giant is closely linked to water and excrement. Hippos spend most of their day in the water in groups of varying size depending on the season. The dominant males mark their territory on land, at the water's edge, by depositing their excrement across a radius of 2 metres. This seems to excite the young, who come to sniff and sometimes consume the territorial markers. The group's structure is rigid. Immature males and females without offspring stick together while mothers with their young gather at a distance from them. The mature males position themselves around the females and each seeks to control the bathing space nearest the female hippos he covets. Each dominant male has his own harem. If the other males and the young adopt an attitude of submission around them, they are not threatened, but if one of them does not stay in his place, the situation can turn ugly. The troublemaker is called to order by postures of intimidation, such as opening the mouth very wide, for example, and charging. The culprit must then submit himself, which he does by generously bespattering the face of the dominant male with his excrement. Every male who enters the water must similarly pay his "respects" to the dominant males. Adolescents males are chased from the group at the age of five, when they regroup in smaller clans in which they remain until they have grown powerful enough to challenge the dominants and acquire their own territories and females. Unlike most mammals who live communally, these giants will not shun dangerous physical contact and will even fight to the death in

overpopulated regions.On land, the hierarchy between dominating and dominated disappears. Piles of excrement are used as sentinels and beacons. Shrubs, termitaries or forks in the river are used as depots for excrement, and the size of the piles that amass in these places can be considerable. They are principally constructed by adult males to demonstrate their predominance. Do these olfactory beacons mark the boundaries of a territory or are they rather visiting cards that enable individuals to recognise one another? The question remains an open one. The one irrefutable fact is that no other animal attributes as many functions to the exercise of nature's needs as the hippopotamus.

Baboons also congregate in very large groups made up of between thirty and eighty individuals of both sexes and all ages. There are around twice as many males as females who are also twice as big and distinguished by their huge canines. They ensure the safety of the weaker members of the group and during daily marches position themselves at the front and back of the troop. These monkeys have a complex social life. Every member of the group knows each other personally. Baboon families, like elephant families, are composed of related females and their young with the oldest female at the head. She determines the group's hierarchy. Adult males move among them more or less at will but the females' tight cohesiveness allows them to impose their wishes on the much more powerful males. Young females begin to mate at five and stay in the group. Young males reach maturity at the same age but must remain chaste for another three years.

48-49 Like wolves, African wild dogs live in packs; a dog separated from his companions will howl with sadness. Nomads, they wander continuously across an immense territory and are very difficult to track.

49 From dawn to dusk a baboon troop travels across its domain in quiet pursuit of food. All baboons are attached to groups. In the evening they gather together in the tree-tops to sleep.

50 top *This buffalo has rubbed his horns against a yellow fever acacia, the bark of which was once ground by the Masai to produce a red dye for their clothing.*

At this point they will join a different group in order to avoid inbreeding.

It takes several months before the new group will accept the young males. Most herbivores live in groups within which certain individuals have a specific role. Gnus and topis provide a good example of this. In May, the migrant gnus of the Serengeti regroup in a tight mass and graze upon the short grasses that they will soon leave behind. Although chaos seems to reign over this multitude, the animals are in fact very well-organised in groups of a dozen to a hundred individuals. Throughout this massive herd some of the gnus can be seen holding their heads up higher than the others; these are the "territorial males". They alone mate, attracting females to their parcels of territory and denying other males access. In a period of only three weeks, 90% of the females - 400,000 animals - are "inseminated".

The dominating gnus must constantly change and redefine their territories as the migrating herd advances. When the herd stops to graze these males impose their authority: stopping females from joining other groups and fending off bachelor males with no territory.

Incessant fighting results: the adversaries face each other on their knees and strike each other violently with their horns. This type of reaction is beneficial to the species, as immature males without territory are prevented from reproducing and kept away from the best grazing grounds. They thus cannot compete for females and their young. Instead, they often fall victim to predators, as their conspicuous movements attract attention, and their constant manoeuvring tires them out.

The topi is also a gregarious animal. Its territories vary depending on the availability of food. Territorial males assemble between three and twenty females and their young. To mark out their territory they deposit excrement and secretions on grass and on the termitaries from which they like to stand watch, guarding against males from bachelor clans. These herbivores migrate in large herds and can be found by the thousand in small, densely occupied areas that are rich in food. The size of the territory the males have to defend varies from small areas measuring 25 metres in diametre to much wider areas covering several acres, where they have much more difficulty maintaining their females.

52 and 53 Bullock-driver birds have strong, sharp-edged claws, a rigid tail and short legs that enable them to keep a grip on the buffaloes which they live upon. Using their narrow, scissor-like beaks they pluck the parasitic insects off the herbivore's back and pick the dead flesh around the animal's wounds, thus cleaning and "disinfecting" them while feeding themselves.

54 *top left The waterbuck is a large animal whose long, reddish-brown coat is made waterproof by a greasy, foul-smelling secretion. There is a long white mark on its hindquarters. It lives in wooded areas near watering holes or marshes.*

54 *top right Antelopes with pale, thick coats, such as the bubal, can withstand the midday sun, while darker-coated antelopes must seek shelter in the shade.*

54-55 *Impalas are among the most graceful antelopes of the savannah, capable of leaping 3 metres high and 11 metres long. The dominant males try to attract and then maintain a harem of females. During the rutting season, fighting between males over territory is common.*

55 *top The horns of young male impalas first appear at the age of two months. These antelopes are weaned at six months and leave their mothers when they are about one year old.*

55 *bottom The oryx is perfectly adapted to semi-desert zones. Its long pointed horns are more than a metre long and an adult can kill a lion with them.*

56 Black rhinos are solitary animals. Bullock-drivers are often their only companions. These birds clean the rhinos' hides of parasites and serve as sentinels while they sleep.

57 top Rhino horns, like the nails and hair of humans, are made of keratin and contain no bone. These pachyderms graze on the branches, leaves and bark of shrubs that they pull to their mouths with their upper prehensile lip.

57 bottom Black rhinos mark their territory with their excrement, which they position by kicking forward with their front legs. The confrontations most commonly observed involve males and females. A female will not let a male approach without putting up a fight. The courtship strategies of a male are hardly romantic: groaning, charging, head and horn butting, pawing the ground, defecating and spraying urine.

58 *A leopard's long canine teeth enable it to kill its prey swiftly and cleanly.*

59 *A female stretches before setting out to survey her domain. She is on heat and her cries and the markings that she will leave are invitations to the many males in the area. She will soon be joined by one of them and they will spend several days hunting together and mating; then they will part company and recommence their solitary lives.*

60 top Leopards are found all over the African continent, except in great desert zones like the Sahara. They are also found in the Middle East, Southern Asia and Indonesia.

60 bottom Leopards stake out the boundaries of their domains, usually by spraying tree trunks with their scent, which will communicate not only their gender but also their age and sexual status. These territorial markings will last throughout the day, thereby limiting the risk of encounters and fights between males.

61 The leopard, with its magnificent coat, elegant body, graceful movements and extraordinary eyes, is one of the world's most beautiful cats. But this beauty has cost thousands of them their lives. A total ban on all leopard skin trade is the only thing that can seriously curb poaching.

62 top Pelicans are very gregarious, and one of the few birds that practises collective fishing: swimming together, they form a circle and plunge their beaks into the water in unison to catch the fish trapped between them. Their large beaks serve as nets but not, as is often assumed, as a means to transport food.

62 bottom left The jabiru is the largest of all African storks. It feeds on fish and frogs, and often throws its quarry in the air before swallowing it. Males have brown eyes, females have yellow.

62 bottom right
The oricou vulture is the largest in Africa, identified by its size, its completely bald head and its beak, which is strong enough to cut into a carcass.

63 Ruppell vultures nest in colonies on cliffs or in gorges. Consequently, they often have to travel distances of up to 100 kilometres to follow migrating herds.

64-65 The mating display of cranes is spectacular. The couple take turns making graceful movements around each other and intermittently leaping up to 2.5 metres into the air.

65 top left The cries of male ostriches on heat cause the tops of their necks to swell as they strut in front of the females. This display is shortly followed by mating.

65 top right The bullock-picker heron lives with cattle and certain other herbivores such as buffaloes and elephants. It feeds on insects and small animals disturbed by the movement of the large herbivores, on which it perches in order to observe its surroundings.

No animal portrait gallery would be complete without a portrayal of the ostrich. Some male ostriches grow to a height of 2.75 metres and can weigh up to 150 kilograms. Though this strange bird cannot fly, it has perfectly adapted itself to life on the plains. The way it advances in great strides of over 3 metres, using its wings to stop and change direction is a sight to behold! Ostriches move in flocks which are dominated by females. They survey their surroundings from the vantage point of their great height. The combination of this strategic position, a well-developed sense of hearing and keen sight make ostriches, like giraffes, the true sentinels of the savannah. For this reason, antelopes like to share their grazing land with ostriches. During rutting season, deep bellowing from males can be heard all night long, coming from their mating territory. They cry in order to attract females to their nesting zone and to warn off other males from their property. During this period, groups and families tend to break up and those old enough to reproduce form groups according to their sex.
Fighting among males can be violent and the wounds inflicted by sharp inner claw can sometimes be serious. When the groups meet up, vividly-coloured males put on spectacular parades of seduction, performing a sort of dance which lasts up to ten minutes. Often, a female will choose a male and begin to strut as well. The ritual complete, the two birds go off and mate. When the population is big, males are polygamous and mate with three or four other females. Each female contributes to the nest, laying six to eight eggs at a rate of one every two days. Only the dominant female, the one first chosen by the male, hatches the eggs and brings up the chicks with the male - the other females only lay the eggs.

THE RULES OF THE GAME

66 top The female leopard spends most of her time with her cubs during the first days of their existence; she holds them against her for warmth and licks them clean. Her saliva dispenses vitamin D when exposed to the sun.

66 bottom Leopard litters number from one to six but most die in the first hours after birth and the mother, desperate for protein, eventually eats the bodies. A new-born leopard cub weighs between 500 and 800 grams. This is relatively small when compared to an adult male, which weighs around 60 kilos, but it will grow quickly.

67 Even though the leopard is mostly a nocturnal animal, the mother carries her young into the sun near their hiding place to warm them. The mother's tail is the cub's first toy. When the female has had enough, she settles down a little way away in a tree where she cannot be disturbed.

When the sun is high in the sky and the savannah dozes under the oppressive heat, female antelopes, gazelles and impalas are among the many animals that withdraw from the rest of the group to give birth. The mother immediately consumes the placenta, then cleans her offspring and remove it from its birthplace as the ground is now marked with odours that might attract predators. After the new-born animal has been well hidden the mother leaves it and returns to the herd. Its natural camouflage is so effective that on several occasions while driving around the great Serengeti plains, we almost ran over baby Thomson gazelles lying totally immobile on the ground. Their coat, darker than an adult's, blends perfectly with the earth! The mother visits the fawn several times a day to feed it. She is careful to carry it from its hiding place first, returning it after it has been fed. The fawn does not join the herd until it can run fast enough to flee with the adults when danger arises, as swift escape is its only protection against its numerous predators which include felines, hyenas, jackals, baboons, eagles and vultures. Gnus behave differently. Gestating females openly assemble in order to keep a better eye on their surroundings. In an emergency, they can even delay giving birth. Babies are born in the morning. During the weeks when births are taking place, the plains are scattered throughout with small calves, hardly five minutes old and still wet, struggling to get to their feet under the vigilant eye of their mothers. The other adults form a barrier around them and watch their first steps. This extraordinary synchronisation of births - nearly 80% of females over the age of three give birth in the same three weeks, generally in January - assure the survival of many of the young. With so much prey at their disposal, predators can eat their fill without wiping out the herd. Few of the remaining twenty percent who are born outside this period never survive.

Young zebras are frequently born at dawn and in their case the whole family is present at the event. As with all young herbivores, the development of the baby zebra is fast. It struggles to its feet and starts to walk behind its mother just two hours after it has been born. In another two hours, it knows how to suckle, trot and even gallop - all essential survival skills. For the first ten days of its life, the other adults of the herd are kept at a distance. The only ones the mother will allow near the baby are a brother or sister. But when danger threatens, the stallion and other female adults join forces to protect it. We have witnessed on a number of occasions adult zebras resist a wild dog attack on a young zebra strong enough to keep up with them. There are few other species, with the exception of elephants in certain circumstances, that can defend a newborn animal against a predator attack. What has more serious implications on the survival of a species is that a female giving birth is especially easy prey for hyenas and lions. This is why female giraffes stay standing during birth despite their great height. One evening in the Masai Mara, we saw a strange-looking giraffe; she was standing shivering at a distance from her companions. It soon became apparent why she looked so odd: two small feet were protruding from her hindquarters! Suddenly, the head of the baby giraffe appeared; the baby slid to the ground, pulled down by its own weight. The amniotic sac broke open and the cord snapped during the 2 metre drop. The baby giraffe lay stretched out on the ground and for the first few minutes it did not have the strength to lift up its head. Its mother licked its whole body with her tongue, giving it strength. After close to an hour it got up on its very long legs, propped itself against its mother's legs and reached up to her teats. By that time, night had fallen. The birth had taken place very late in the afternoon, with lions in the vicinity.

68 Among giraffes, tender scenes between mother and infant are rare. Giraffes' maternal instincts do not appear to be very developed, and mothers will not hesitate to abandon their young while they go off to feed. In the Serengeti, abandoned baby giraffes are often picked up and reared in a sort of animal nursery.

69 At birth, the baby giraffe measures nearly 2 metres and weighs about 60 kilos and for the first few months of its life it grows 3 centimetres a day! It does not need to feed regularly as its mother's milk is rich in nutrients.

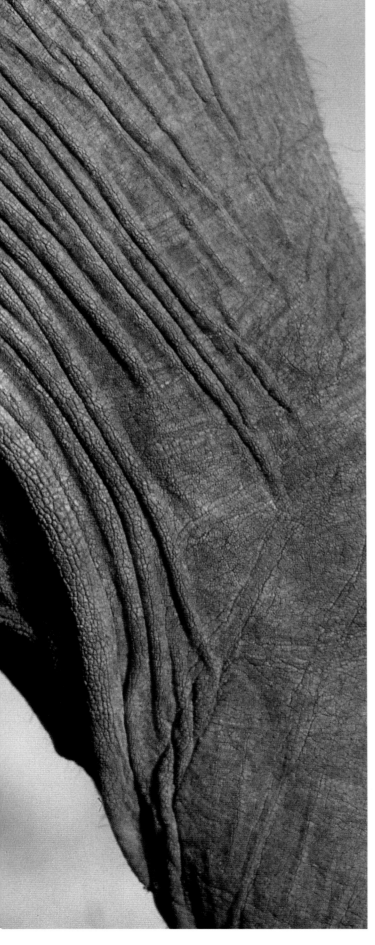

Two marauding lionesses soon arrived on the scene. The mother put herself between the lionesses and her baby and stamped her hooves against the ground in hard, resounding blows. But one of the lionesses managed to swat the baby, who fell down and did not get up again. The mother spent an hour trying to defend it and then, appearing suddenly disinterested, moved away to eat. The two felines helped themselves to their kill.

Herbivores are not the only newborn animals that are easy prey for predators. Baby felines and hyenas can also fall victim to predators, but have more to fear from adult males of their own kind. A mother will keep them hidden for the first few weeks after their birth and, in order for them not to be tracked down by their odour, she will frequently carry them one by one in her mouth to a new hiding place. Adults do not only have to take care of the safety and survival of their progeny, but also see to their education. Raising young is a difficult task, especially when the mother must do it alone, as is the case with cheetahs and leopards. Male cheetahs and leopards only join females to mate, after which the female resumes her solitary life. When she becomes a mother, she has no choice but to leave her young alone and at risk while she goes off to hunt. Cheetah cubs are merely hidden in the hollow of a ditch or in a bush, where, for example, they could easily be crushed by a passing herd of buffalo. When the young are ready for solid food, their mother must then not only provide for her own needs but theirs as well. A female cheetah with a litter must hunt everyday, while without offspring she only needs to hunt every two or three days.

Communal species raise and defend their young in a completely different way.

70 top A baby elephant is constantly reassured by the sniffing and stroking of the trunk of an adult. It will be three or four months before it starts to use its own trunk.

70-71 A baby elephant searches for its mother's teats with its trunk which at this stage it does not use for anything else, as it drinks through its mouth. Moreover, it has to bend it over its head because it impedes the sucking.

72 These two baby elephants have just gone to sleep, protected by the wall formed by an adult elephant's legs.
As soon as they wake up they begin to play with each other or with a plant or branch. They can be entertained by very little. Intensely curious, they are fascinated by all the animals they see: birds, tortoises and baboons. They pretend to charge, then either proud of or frightened by their audacity, they go back to hide behind their mother.

72-73 An elephant is born with black or red hairs on its head, forehead and back. These gradually disappear as it grows up. At birth, it weighs nearly 120 kilos.

73

74 top These adolescent male hippos are being initiated into the social behaviour of their future adult life. They become sexually mature at thirteen and so become potential rivals of the dominant male. This is when they will be chased out of the herd.

74 bottom A female hippopotamus is minding the nursery. The other mothers have left the water to graze on land. The young hippopotamus will be able to leave the water to graze in about six to eight months.

74-75 Just five minutes after being born, a baby hippopotamus can already swim and walk. It suckles most often underwater, holding its breath with its nostrils and ears shut. Its mother stays with it for the first few days during which time she is very aggressive.

75 top *A young hippopotamus always stays close to its mother's head, where she can best keep an eye on it.*

Though the female hippo gives birth away from the herd, both mother and infant return to the herd after ten or twelve days. They rejoin the group which is made up of other mothers and their young ones, far away from the males, who may attack or trample on them by accident. The female hippo ferociously protects her young, particularly from the males, but she does not always succeed. When a dominant male conquers new territory, he also wins the right to new companions. If these females are feeding their young, this stops them from being on heat. The new master may sometimes decide to kill all the young who are not his own so that the females will mate with him more quickly.

This infanticide is similarly practised by male lions. During the day, one or two lionesses watch over all the lion cubs on the bank of a river, forming a kind of nursery so that each mother can take it in turns to find food. The female lion community is so close-knit that the cubs can suckle any mother. This is a unique phenomenon among felines, and extremely rare throughout the animal kingdom. Most baby animals that have the misfortune to lose their mother are left to die, since no other female will feed and take care of them. It is easy for lionesses to put their resources at the disposal of the community as they often go on heat and give birth at the same time. One of the loveliest sights to behold on the Savannah is a group of lion cubs, which can number as many as twenty, aged between six weeks and five months throwing themselves on their mothers returning from the hunt followed by their companions. There is a great deal of meowing, squealing and scratching as the hungry cubs fight to get to one of the teats.

Female elephants and hyenas also bring up their young collectively. Among wild dogs

and mongooses, the entire herd, males and females alike, take care of the young.

We camped next to the den of a pack of wild dogs in the middle of an immense plain in the Serengeti for several weeks. It was a big litter of eleven pups. Feeding times were chaotic as the pups squirmed and grappled, trying to lay claim to one of the teats. When the adults went off hunting, the mother stayed with them in the den, sometimes accompanied by one or two of the males. Once they had grown up a little, the adults in the pack shared the duty of watching over them. Upon their return, the hunters would regurgitate pieces of meat for those who had stayed in the den.

The adults are very affectionate to pups and will take care of all of them, without showing any preferences. While looking after a pup, they will playfully bite it, lick it, roll it around on the ground and then go on to the next one, who yelps in eager anticipation of his turn. At the slightest cry of alarm, the pups run back into the burrow. Because of the time they dedicate to raising their young, neither wild dogs nor mongooses produce litters at close intervals which is why only the dominant couple reproduce.

76 While at rest, the female cheetah lets her young ones tease and playfully bite her but will call them back to order with a swat of her paw if they get out of line. After five weeks, they follow her on to the plain, highly visible to predators and still too young to flee quickly. A dangerous age!

77 This six week-old cheetah cub is not yet ready to follow its mother into the wild. She calls him when she goes off to hunt or she brings him back her prey still intact, dragging it in her mouth. Young cheetahs always eat first.

The birth rate of animals is also controlled in accordance with the availability of food. In periods of drought, for example, the average birth rate of elephants drops. Over the course of our many trips through the savannah, we have witnessed, much to our delight, the many different games elephants, lions, cheetahs and monkeys play among themselves.

How we have laughed watching young baboons swinging from each others' tails doing somersaults and leaping from branch to branch. Animal games take on various forms. We have seen a cheetah cub trundling elephant dung, a young leopard turning over a tortoise, or a young jackal, fresh out of its lair, playing with anything it can find - dung, blades of grass, feathers, butterflies, frogs or a sibling's tail. There is no end of pleasure to be had watching lion cubs chasing each other around frantically and learning about their surroundings through play, and there is only one way to find out that a porcupine does not make a good playmate! Unusual objects, such as our vehicle, did not seem to frighten the young cheetahs we had been following around for several weeks. We always had to have a good look around before setting off as they had a habit of running around and hiding underneath it. They even discovered the malicious pleasure of gnawing on our brake cables! Defining the game is difficult, even for those who study animal behaviour. But for whatever reason, the young spend an enormous amount of energy when they play and their recklessness sometimes leads them into danger. Distracted, they may stray from their mother without even realising and predators are never far away. But the benefits of play are obvious. It helps the young to build strength and develop agility. Little Thomson gazelles somersault and bound with disconcerting adroitness the minute they come out of hiding. They quickly acquire the skills they will rely on throughout their lives to escape from predators.

78 top left These three month-old cheetahs have shed their first coat - just vestiges of the superb dorsal tail remain - and are playing more and more energetically. This family of three are chasing each other around and setting up ambushes by hiding behind a tree trunk or a rock. Through these games they are learning how to hunt.

78 top right With its long legs, body and tail, and deep-set chest, the cheetah's streamlined anatomy is perfectly designed for running.

78-79 The female cheetah is carrying her five day-old cub. She usually produces a litter of three or four. At birth, cheetah cubs are snow-white with black bellies and weigh less than 300 grammes.

80 *As soon as they are developed enough, young cheetahs begin to climb small acacia trees; their non-retractile claws are not yet blunt, enabling them to perform acrobatic stunts that adults no longer can.*

81 *How quickly this nine month-old cheetah will mature depends upon its sex. The female leaves its mother to settle down alone in a territory. If there are a few of them, the males stay together and form small groups. Barely one out of three cheetah cubs survive to reach adulthood. Their mortality rate is sometimes as high as 70%.*

Animals learn about life on the playing field. Through games they learn stealth and cunning, skills they will need when the time comes for them to acquire territory and hunt. They also learn how to be on their guard and how best to react to unexpected circumstances.

The type of games played reflects the society in which these young animals live. They copy their parents' gestures and attitudes. Young baboons learn how to recognise members of their community and understand the relationships which exist between their different groupings. Wild young pups and hyenas share their discoveries, like a branch or a piece of bone. When they fight, they are learning the rules of hierarchy that govern the pack. The sex of the animal is also reflected in game playing. Young male baboons and elephants fight more than females of the same age. Fighting between young males enables them year after year to establish a hierarchy among their peers. By the time they are adults, males will know what their social position is in relation to others. The same phenomenon exists among zebras. Adolescent stallions form bachelor groups that grow up together and gradually learn through all kinds of games about the strengths and weaknesses of their future rivals, which they will profit from when the time comes to start a family. At first, as when they were infants, they rear, lash out, try to bite each other on the nape of the neck, throat and legs, while dodging flying hooves and their adversaries' bites in the course of short, fast races and chases. The hierarchy that will govern their relationships gradually emerges.

Young male monkeys however, only think of chasing after each other, while their female companions take obvious pleasure in handling and delousing the babies.

83 top The young black rhino is born without horns; they begin to grow at about the age of two or three. The mother and her new-born stay hidden in the bushes for the first two weeks when the mother becomes quite irascible. For the next two years, the young rhinoceros follows her around everywhere, after which it is time to be weaned. It is now so big that it must lie down to reach the teats.

83 bottom In the days when the rhinoceros population was large, adolescent black rhinos would gather in corresponding age groups. Because of intense poaching, however, their numbers have been severely decimated and they can no longer be so discriminating

Female elephants also take better care of their young than their brothers and male cousins. They help the young ones get over obstacles and, like their mothers, reassure them with their trunks.

Games are indispensable for yet another reason. For animals as well as humans, they channel violence. The stronger animals play with the weaker ones at their level. Adolescent elephants kneel down so that younger elephants can climb over them. The older lion cubs of a pride never deploy all their power when faced with a weaker adversary; by restraining itself from winning, the stronger animal makes the game last longer. Big lion cubs tussling on their hind legs look like they are fighting ferociously. But, all of a sudden, one of them will stretch out and expose the most tender part of its anatomy to its rival, who, of course, will not take advantage of it: in growing up and playing games the animal has learned to recognise the rituals and rules of life in society. During the jousting, it is obvious by their expression that the participants are only play-fighting. Their mouths are open and relaxed, teeth half-bared, signalling good humour and pleasure. Each animal shows the other its friendly intentions.

Canines and felines add specific body language to these signals: curved back, flexed forepaws, hindquarters up in the air and vigorously waving tails.

Not all animal species have the same relationship to games. The less evolved they are, the less their young ones have to learn to attain maturity and the less they play. Even for the most evolved species, species which play the most, games never take place during potentially life-threatening situations, namely, during eating and reproducing. The elephants of the Amboseli reserve provide a good illustration of this phenomenon.

84 top These young hyenas were born of two different mothers from the same den, but neither will venture out unless it hears its own mother calling, whose voice it can easily identify.

84 middle Young wild dogs begin to eat meat when they are 15 days old, even though they will suckle for two months. In order to provoke the regurgitation of the food, they lick the corner of the adult's lips.

84 bottom The female hyena is carrying her several day-old baby to another den. It was born black with its eyes open and was already quite developed. The spots on its coat will appear in about a month and a half when its coat lightens.

85 This young spotted hyena still only feeds on milk. It will be milk-fed until it is one year, perhaps even 18 months old. Adult hyenas do not regurgitate meat to nourish their young. By the time it is weaned, this pup will have almost reached adult size.

Food is scarce at the height of the dry season. The Amboseli elephants live in small groups of about ten and spend most of their time foraging to satisfy their needs.

They therefore ration their strength and the young elephants do not play. During this period they seem almost to droop with sadness.

Once the rains have come, however, and everything has turned green the difference in attitude and behaviour is marked. The groups join together to form bigger units and the happy games are renewed with vigour. Whatever form these games may take, scientists are sure that they are of great importance.

Recent research on the brain shows that games are perhaps as important in life - both for humans and certain animals - as sleeping and dreaming. Proof of this is that the absence of games among young animals, as well as human infants, brings about serious pathologies.

Studies of monkey behaviour, and in particular, chimpanzees, show that animals deprived of games in their youth (and this is true of animals both in captivity and in the wild) are immature, introverted and shy, and incapable of socially adapting. Animals born in captivity and raised alone cannot communicate properly with others when they are adults.

They cannot situate themselves in a hierarchy and can neither woo nor mate with females. Games are therefore important and indispensable in the life of young animals. They encourage good physical development and facilitate sociability. But games should not be reduced to just that; perhaps animals play for the sheer pleasure of playing.

Games are not only for the young. Adults, too, like to play, though they spend much less time doing so.

88 This lion cub of less than six weeks and its brothers have not yet joined the rest of the pride. Soon, they will begin to eat their first solid food. Their mother is very aggressive to anyone who comes close to her children.

89 The female lion transports her young one by one by delicately gripping the scruff of their neck between her teeth. The new hiding place may be a ditch, the hollow of a rock or simply a sheltering thicket.

90 Nearly three-quarters of all lion cubs die before reaching maturity. The negligence of their mothers is one of the causes as they have a tendency to move on without waiting for their young. They are preyed upon by hyenas and can also be killed by male lions. When nomadic males conquer those who are territorialised, they generally kill all the lion cubs so that they can procreate more quickly. Lion cubs also die of hunger when prey is rare and adults invariably eat first.

90-91 This young lion cub is playing with a weaver-bird's nest that has fallen out of an acacia.

91 top These tiny lion cubs look like balls of wool. Their future depends on their gender. Females stay in the pride while their brothers and male cousins are chased out after two or three years. They generally stay together, which increases their efficiency in hunting and conquering territory after several years of wandering.

92 top The lioness is annoyed by the cubs who have pulled too hard on her teats. She pulls back and bares her teeth to those who want to feed again.

92-93 Though the male lions of the pride tolerate the games of young cubs, they are less indulgent than the females. They spend most of their time together, away from the rest of the pride.

93 top It is not the lion cub which has brought about this male's reaction but the urine deposited by a female several moments before. The male sniffs it, his chops turned up, in a characteristic grin.

93 middle Lion cubs are weaned at six months but many continue to suckle until they are a year old, and sometimes longer, often depriving other new-born cubs of nourishment.

93 bottom Unlike male lions, adult females often play like cubs. Their games intensify in the rainy season but die down at the end of the dry season when food becomes scarce. Female lions are also very affectionate with each other.

94-95 The young zebra suckles at its mother's teat for up to six months. Then it continues to graze in her company until it is two years old. If it is a female, she will leave the family group and wait to be chosen by a stallion and join a new family. If it is a male, he will join a group of young bachelor stallions for one or two years before attempting to establish his own family at the age of four or five.

WATER: THE SOURCE OF LIFE

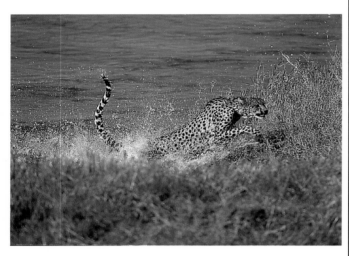

96 top Cheetahs do not like water but they can cross a shallow river if they need to.

96 bottom Cheetahs who live in arid zones reduce their water consumption to a minimum. They can survive by drinking their victims' urine and eating desert melons.

Rain and the absence of rain determines life in the savannahs and can also completely change the landscape. At the height of the dry season, the ground is cracked and dry, the grass is yellow and the animals thin and weak. Then the rains come, setting in motion the whole ecological cycle. Thousands of organisms which, having taken refuge from the heat in the dried-up ponds and lain dormant for months, reawaken. A myriad of insects come out, the first batches of the year, and are greeted with joy by the flocks of birds and predators who suddenly find themselves surrounded by food. The grass quickly becomes verdant, dirt turns into mud and the swamplands become, once again, impenetrable. Only elephants, buffaloes and waterbucks can safely venture in search of grazing grounds. Food becomes easy to find again. The rainy season is a period of abundance for both herbivores and predators; birth rates peak and migratory species return to the plains. Games start up anew. The most striking phenomenon linked to the rainy season is the migration of two million herbivores. In November, over a million and a half gnus, as well as thousands of zebras and Thomson gazelles converge upon the Serengeti plains. The short, green grass which covers these vast, empty stretches of land will feed a multitude of animals until the month of May. When one sees such a fantastic concentration of animals, it is hard to imagine that in just a few months only a few gazelles will left living on the dried grasses of the plains. In May the animals begin their march, as if impelled by irresistible forces. The lowing herds slowly make their way northward.

These interminable lines of animals on the move, which cover up to 40 kilometres without a break, leave deep furrows in their wake. Every so often they break out running in a sudden frenzy, and then, for no apparent reason, just as suddenly they stop.

96-97 Leopards often choose to live near watering holes but they can only drink every two or three days. They like water and are strong swimmers.

97 top This female cheetah and her young cub waited for the rain to end without moving. The mother soaks one of her young ones while shaking herself off.

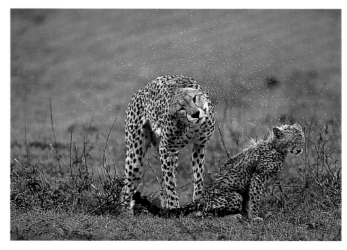

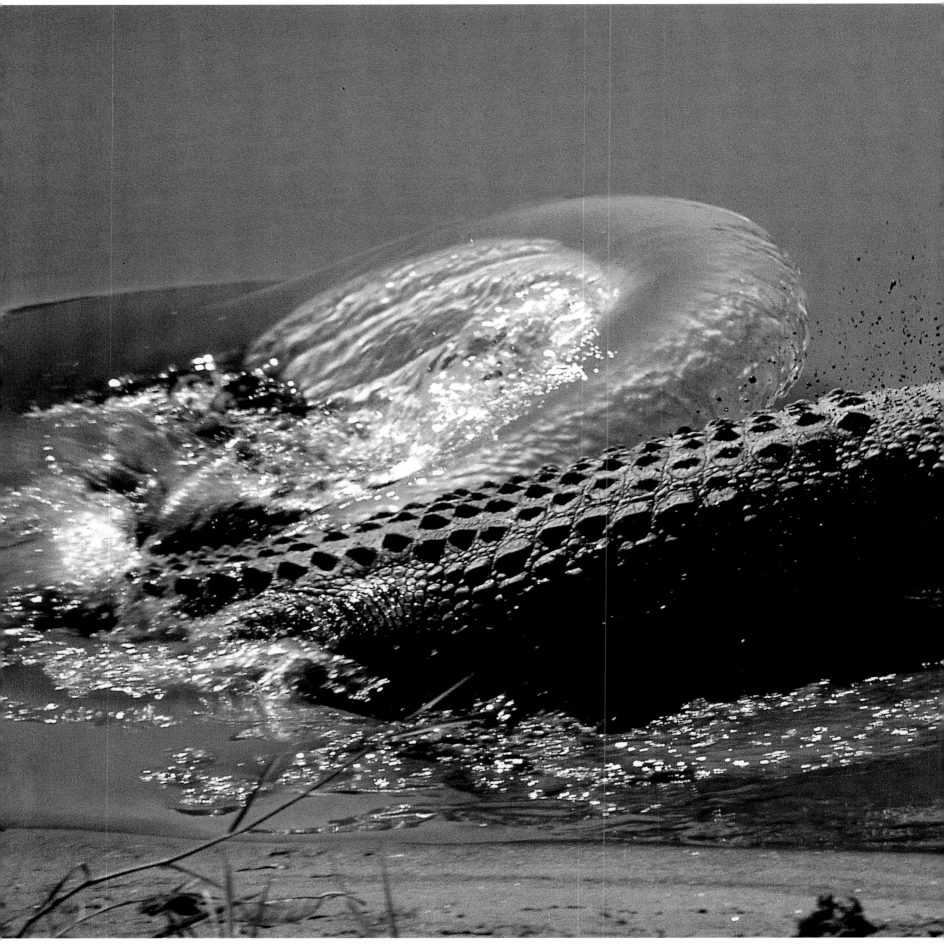

98 top left To drink, giraffes stand with their front legs wide apart so that they can stoop down to reach the water. This posture puts them in a very vulnerable position. In lowering and raising themselves, the amplitude of the movement of the giraffe's

brain puts an enormous strain on its heart. The giraffe's blood vessels, however, are highly elastic and the veins in the neck are equipped with special valves. Its heart, which weighs over 11 kilos, pumps out approximately 60 litres of blood per minute.

98 top right Zebras need to drink at least once a day during the dry season. They are always watchful at the watering hole. If one of them senses approaching danger, he warns the others by vibrating his nostrils, a signal which carries over a long distance.

98-99 A crocodile's tail constitutes 40% of its total length. It is an excellent instrument of propulsion.

The huge mass of animals divides into two groups: one heads northward to the Mara river, crossing the border of the Serengeti when it reaches the Masai Mara reserve; the second group turns westward, taking what is called the "Western Corridor" towards the abundant grasses of Lake Victoria. This exodus, which covers a distance of nearly 1500 kilometres there and back, is strewn with hazards. It is a terrible test for the species and a windfall for predators. The mortality rate of young gnus is shockingly high. Young gnus who lose their mother in the general commotion are condemned to die. They are easily preyed upon by predators or are victims of fatal accidents. When summer is over, the gnus go back the same way they came. Among the most difficult obstacles they must face on the way are the river crossings, particularly over the Mara. The gnus invariably cross it at the same specific points. These spots were once easy to cross before but soil erosion and repeated stampeding year after year have transformed the river banks into small cliffs. Many of the animals who converge en masse on these steep precipices either drown or are trampled. The river is quickly cleaned of its carcasses by the resident vultures, crocodiles, monitor lizards and catfish who eat everything that comes their way. Migration is linked to food needs, the perpetual quest for greener pastures and water. As they are dependent on the rains their dates and form change from year to year. Nevertheless, the history of migration is as old as mankind; the Olduvai Gorge fossils (where a 1,750,000 year old Australopithecus was found) at the gateway to the Serengeti provides evidence that the gnu grazed here a million years ago. Not all animals have the same water requirements. The oryx, dik-dik, Grant gazelle and giraffe gazelle can get by on very little, as they aborb enough water from their food. Buffalo and elephants, on the other hand, need a lot of water which excludes them from living in certain areas. Buffalo must drink between 30 and 40 litres of water a day and elephants, between 70 and 100 litres.

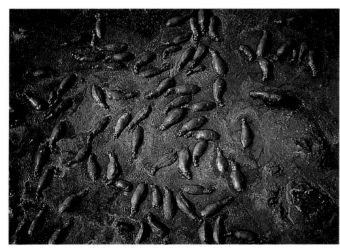

100 top left Hippos love the mud. Certain extremely muddy swamps are colonised only by dominated males. They can spend all day wallowing in these confined spaces, occasionally rolling around to sink their backs into the mud.

100 top right The hippos are surrounded by a kind of foam which spreads along the surface of the swamp; it comes from the fermentation of organic matter expelled by animals.

100-101 and 101 top Hippos do not suffer from insect attacks as much as other large mammals; they can avoid them by burying themselves in the mud. When they are above the surface of the mud, their bodies are meticulously inspected and picked over for parasites by jacanas and bullock-driver birds.

101 bottom In shallow water hippos move in leaps and bounds, bouncing off the bottom.

102-103 This male hippopotamus has come out of the swamp but wants to go back. In order to do this he adopts a submissive attitude towards the dominant males whom he may disturb.

Both these animals, along with the hippopotamus and the warthog, adore wallowing in the mud which dries rapidly and is refreshing. The mud suffocates parasites lodged under the skin and stops horse-flies and flies from laying their eggs on them, providing a vital and necessary form of hygiene.

Though the second largest land-based mammal, the hippopotamus is completely at home in an aquatic environment and spends most of its time in water. This is the cause of its corpulence and unusual shape. Not only does water account for a part of its weight but it plays an essential role in its thermoregulation.

A hippopotamus dehydrates three to five times faster than a human being in the sun; consequently, it cannot stay in the sun for long periods of time. Its skin is surprisingly delicate and under the sun it produces a vivid red and viscous liquid which makes the animal's back look like it is bleeding. This perhaps explains why in Egyptian hieroglyphics the word "surgery" is represented by a hippopotamus!

The so-called "river horse" is well-adapted to its environment: its eyes, nostrils and ears, which are shut by valves at the top of its head, enable it to observe without being seen. It can hold its breath underwater for five to eight minutes. The waste matter expelled by these gigantic creatures provide tons and tons of rich manure, allowing for the proliferation of phytoplankton and all types of aquatic plants, which a multitude of small crustaceans, larvae, worms and other invertebrates feed off. This is why so many fish-eating birds hover around these mammals. The dispersion of dung by the hippo's tail has enormous repercussions on the environment. The heavy excrement sinks providing food for fish.

The lakes of the Rift Valley have formed a unique and extremely diverse aquatic environment. This is the result of the conditions that were present when they were being formed. Most of these lakes first appeared after earthquakes and volcanic eruptions and so many of them have high alkaline levels and are rich in minerals from the lava. As the water is confined, its mineral content becomes increasingly concentrated, due to intense evaporation and the banks are bleached white with salt.

Few living creatures can survive in such a caustic chemical stew. Blue Spirulina algae, however, are perfectly adapted to this environment and reproduce at a fantastic rate. They give the water its blue-green colour, or, in the case of the algae in Natron and Magadi, a deep red. The algae are consumed by flamingos of which there are about three million in East Africa.

104 and 105 Elephants love to splash around in the mud and wet sand. They scoop it up in the hollow of their trunk and smear it on their chest, back, flanks and head.

When it dries, the mud forms a protective crust against the sun and insect bites. Playing in the mud is not only useful, it is highly pleasurable!

Less than a hundred thousand of these are pink flamingos. These birds do not compete with the rest, however, as they feed primarily on arthropods. When searching for food, flamingos walk with their heads down so that the upper part of the beak is horizontal. Their beaks are particularly adept tools, the edges of which consist of thin, rigid plates which filter out particles in the water. Mating rituals occur in groups, often outside the nesting areas and do not always end in reproduction. Dwarf flamingos usually only reproduce on lake Natron in Tanzania. Egg-laying occurs almost simultaneously throughout the whole colony.

The nests, truncated cones made of densely packed mud, feathers and stones, are built close to one another.

The parents mark out a small territory around their nest, which they defend by pecking. The chicks leave the nest after eight days, forming enormous flocks.

Each parent recognises its young by their voices and feeds them by regurgitating food directly into their beaks.

Not all the lakes have such a high concentration of salt, however.

Some are only slightly briny or have entirely fresh water. Aquatic life in these lakes is extremely diverse. The Naivasha, Baringo and Turkana, for example, are renowned for their abundance of fish, crocodiles and bird varieties.

The large number of bird species which thrive in the African aquatic environment is due to the fact that each one exploits a different ecological niche. Pelicans and cormorants are well-equipped to fish both above and below the water's surface. Others, like the kingfisher, only stay a fraction of a second underwater, just the time needed to catch a fish and bring it up. Harpooners, holding on to the ground with their strong, high legs, stretch out their long necks and spear the fish swimming by.

107 top left Elephants suck up water with their trunks and then spit it out through their mouth. When they cannot find any water they use their tusks to sink wells into dried up riverbeds.

107 top right Elephants of all ages love to wallow in mud, which is good for their skin.

106 Bathing is recreational for young elephants. They jostle and climb over each other. Adults also take part in the aquatic gymnastics. A herd will walk over 30 kilometres during the dry season just to take a bath.

106-107 Elephants do not let bullock-drivers remove their parasites, though they are often surrounded by bullock-picker herons. These birds live in flocks; they return to their nests every evening to sleep and reassemble in the same groups the next morning.

108 and 109 Dwarf flamingos have no shortage of enemies: spotted hyenas, jackals, eagles, vultures and marabous. When anxious, these birds draw themselves up into an erect position in a single sharp movement. If the danger persists, the entire colony moves in the same direction. It is an extraordinary sight to see, like an enormous wave that at first breaks slowly and then accelerates, cresting in the midst of great clamour. Ostriches, however, remain indifferent to them.

Herons fish in the same manner but even these specialists divide the different water levels to avoid competing against each other. The Rift valley, with its string of lakes and diverse biotopes, is one of the most singular and remarkable ornithological communities in the world and constitutes an extremely important migratory route, sheltering millions of wintering birds from Europe each year. The most awesome predators in the water are the Nile perch and tigerfish, whose power in the aquatic environment is comparable to that of the great carnivores on dry land. Other fish predators, apart from the considerable numbers of birds, include the Nile crocodile.

The diet of a fully-grown crocodile mainly consists of fish, but it will also attack mammals who come to the shore to drink, and occasionally it will scavenge. An adult crocodile eats on average just fifty times a year storing 60 % of its food in the form of fat. These reserves are sufficient to enable growth which is continuous.

The diet of a crocodile changes with age. At birth, young crocodiles essentially consume insects, which, until they reach the length of 1.5 metres, constitute 30 % of their diet. Between 1.5 and 3.5 metres, they eat molluscs and fish. When they exceed 3.5 metres they start to consume bigger prey. An old adult crocodile can go for over two years without eating and a newly-born crocodile, four months.

By its nature, the crocodile diet keeps the number of Nile perch, tigerfish and catfish down.

As they are also cannibals, adults frequently swallow newborn babies and a large number of their eggs is devoured by Nile lizards and marabous, thus ensuring that the crocodile population never reaches levels that would jeopardise the other populations of the aquatic ecosystem.

110 top and 111 The tantalus is an African stork. It feeds itself by wading through calm waters which it trawls with its beak. As soon as it comes into contact with its prey, the bird stretches out its neck and snaps it up with its beak — one of the fastest movements in the animal kingdom.

110 bottom The Egyptian goose is a duck with a characteristically low and strident cry. During the mating season, females without partners attack other females to take away their male partners.

112-113 During the annual migration vast numbers of gnus head for the Mara river. The first ones to reach its banks hesitate before crossing. Amidst the chaos they panic and try to change direction. With the wave of new arrivals many are trampled upon, crushed against the steep banks, suffocated or pushed into the water. The air rings with the sound of their piercing cries. The young gnus, separated from their mothers, call out in anguish. They will die, as no other female will take care of them.

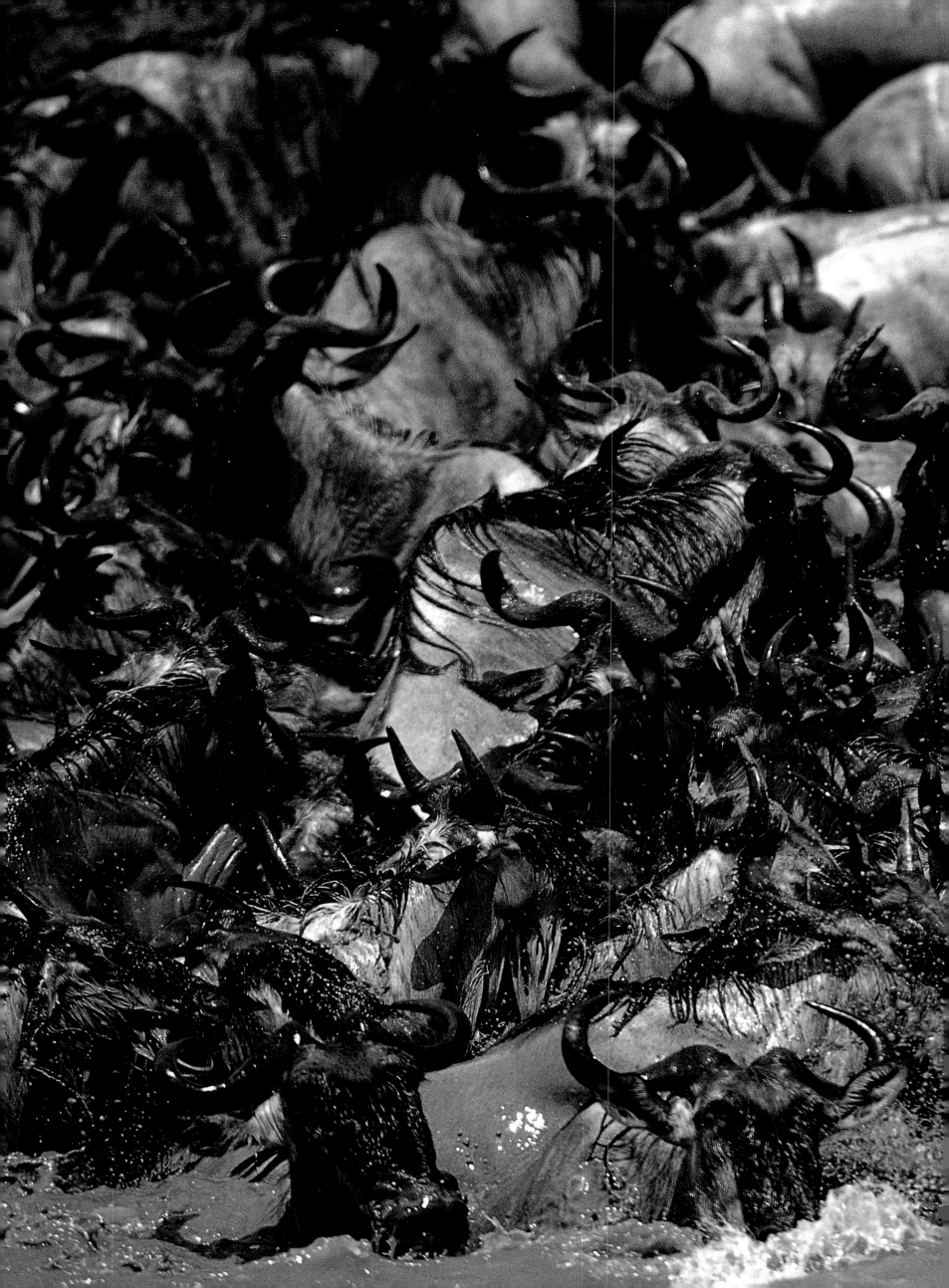

THE FOOD CHAIN

The image of wild Africa has, for a long time been that of an immense virgin forest, mysterious and impenetrable, inhabited by cruel carnivores who mercilessly slaughter herbivores, nature's eternal victims, and even eat each other. In the savannahs of East Africa, zebras, gnus, gazelles and other herbivores quietly graze during the hot hours of the day, often within range of lions relaxing in the shade of small bushes. Leopards, despite their reputation as ferocious predators, spend the afternoon sleeping quietly in a tree. Only the sentinels scrutinize the horizon. The herds calmly graze, stirring only when attacked. Herbivores cannot live in a state of constant stress, as this would prevent them from reproducing. Survival of the fittest is the law of the land in these great African savannahs, which sustain great concentrations of animal life. Plant-eaters are the most numerous and diversified. But all these animals put together only represent one percent of all living matter; the other 99% is made up of plants. They capture the physical energy of the sun through photosynthesis and convert it into chemical energy in the form of organic matter. Vegetarian animals eat their fill before being eaten themselves by predators. Each consumes the one who precedes them in the long food chain, at the top of which are the super predators. Scavengers feed on carcasses whose mineral elements return to the soil, which plants in turn draw nutrients from, bringing the process full circle. Out on the open savannah, all these interactions are easy to observe. The processes of the food chain, like the establishment of animal hierarchy, are daily events that occur in full view. Dung-beetles, for example recycle ungulate excrement by rolling it into a ball in which it lays its eggs and which it then buries.

There are numerous plant-eating animals that can coexist because although they share the same habitat they have different diets. Elands and kudus like forest groves where the vegetation is relatively dense. Oryxes live in dry forest and arid regions, while waterbucks always live near water. Gazelles frequent open habitats in fairly dry zones. An important dietary distinction can be made between those species that eat leaves and shoots and those that graze on grass. Those belonging to the first category have elongated jaws which they work with delicacy while herbivores have wide jaws and incisors like crude pincers and are much less discriminating. Not all animals fall neatly into one category or the other: impalas, elands and elephants, for example, fit into both categories. They eat grass during the rainy season and leaves during the dry season. Leaves contain two to three times more protein than grass, even during the rainy season. Ungulates can also be divided into ruminants and non-ruminants. Ruminants swallow their food rapidly without masticating. It is stored in the stomach where it is attacked by micro-organisms. During digestion important proteins and sugars are produced.

On the other hand, the less efficient digestive processing of non-ruminants necessitates the ingestion of greater quantities of food than ruminants of the same weight. These differences also have an effect on the geographical distribution of the species. Ruminants cannot consume plants high in tannins and resins as they destroy the micro-organisms in their stomach. Zebras, rhinos and elephants - non-ruminants - can survive in dry environments where food is plentiful, but poor in quality.

The size of a species is another important factor that determines the geographical distribution of ungulates and the type

116 top left This elephant is pulling off the bark of an acacia tree which, unfortunately, will die as a result.

116 top right Despite their abundant needs, elephants are discriminating eaters. When the vegetation underfoot has been consumed, they use their powerful trunks to pull down the more succulent and tender branches from higher up. If they are still hungry, some - usually the males - stand up on their hind legs to be able to reach even higher.

116-117 An elephant's diet is varied: grass during the rainy season; leaves, roots, tubers, bark and even wood during the rest of the year. Both the upper and lower jaws have only a single giant, jagged tooth, which is replaced once it has worn away.

of food they consume. Giraffes have access to resources that others do not. Their basic diet is composed of young acacia shoots. As each tree produces only a few shoots at a time, a giraffe never spends too long nibbling at the same tree and consequently the tree itself suffers little damage. When the terminal shoot is severed, a lateral bud takes its place and produces a new shoot. Over several months the giraffe is provided with a store of convenient food, which it takes care. The giraffe eats very little because it chooses the most concentrated food source available, unlike the elephant, which is wasteful.

The elephant needs between 330 and 440 pounds of food each day and it often needs to spend over sixteen hours a day to find it. Big plant-eating species must absorb more food than little ones. As their metabolism is weaker, they need less food per unit mass. Their feeding time being limited, they must eat large quantities of food quickly, without too much regard for quality: bad quality grass for buffalo, stalks for zebras, wood for elephants. Smaller species, with their higher metabolism, digest more rapidly. For them the key factor is the level of protein in their food; the smaller they are the greater the quality of food ingested must be. This is reflected in the choice of plants as well as in the part which is consumed. In the case of migrating animals, large species like buffalo and zebra prepare for the arrival of their young ones. They eat the toughest parts of the grass and leave the most tender parts for the young ruminators - gnus, topis and Thomson gazelles. The gnu, while grazing on grass, stimulates the formation of young, protein-rich leaves which are eaten by small gazelles. The shoots would not appear without grazing. This is why the savannahs of the Serengeti/Masai Mara ecosystems can feed such an incredible quantity of animals.

118 This lioness is strangling a topi, which has been anaesthetised by the shock - a psychic mechanism plunges the attacked animal into a coma-like state, which explains why even if it has not been wounded it will not struggle to get away. A zoologist, wanting to cut a meat sample from an apparently dead gnu which had been caught by lions, was surprised to see it suddenly get up and run when the lions moved away.

118-119 A lioness selects a topi from the middle of a herd and charges directly at it, brushing past another herbivore, which she ignores. Once in full stride, it is impossible for her to change direction. The topi jumps in an attempt to escape but the feline manages to grab its hindquarters in mid-leap, determined not to let it get away.

All large herbivores are valuable prey to the predators which roam the savannah: lions, leopards, cheetahs, hyenas, and wild dogs. These carnivores prefer to attack animals that are either young or old, sick or wounded rather than fully-grown animals in full possession of their capacities. But they have different hunting techniques and diets - some are both predators and scavengers. In the wide open spaces that they generally inhabit, lions are easy to spot. They therefore hunt at night, dawn or twilight. Some prey can aptly defend themselves against a single hunter by coming at it with hooves, horns or, in the case of larger prey, by charging at it. Lions who hunt in groups, therefore, have greater success - a lion by itself has only an 8% chance of killing its fleeing prey, whereas the success rate goes up to 30% if there are more than two attackers. Group hunts enable lions to go after bigger prey. In groups, lionesses take care of 80 to 90% of the hunt, with males participating only minimally. A lion's fastest speed is only thirty-five miles per hour compared to the fifty-five miles per hour capabilities of some of its prey. A lion by itself in tall grass is on the lookout; it locates a potential prey and tries to creep up to within sisty to one hundred feet from it without being noticed. Once it has covered this distance it launches its attack and pounces. With all its weight bearing down on its victim, it seizes it by the throat and kills it. When lionesses hunt in groups they surround their potential prey. Some openly walk up to the prey while the rest crouch and wait. The first ones force the victim to flee in the direction of those hidden in waiting. In the Serengeti, gnus represent close to one half of all lions' prey, followed by zebras, buffaloes, bubals, and gazelles. One adult lion kills an average of 20 large herbivores a year. Dead animals are part of their diet as well and if there is no game, this feline can live off rats, fruit and ostrich eggs.

120 Male lions of a pride rarely hunt. They much prefer to eat animals killed by the females, which they do 75% of the time. Roughly 12% of their prey is stolen from other carnivores who themselves, kill only 13% of the time. Solitary males must, of course, hunt much more frequently.

121 This lioness has just caught a baby impala that was left all alone on the plain. This crafty feline takes advantage of any wounded or defenceless animal.

122 Lions, both male and female, can devour 30 kilos (14 pounds) of meat in one sitting. It is not surprising, therefore, that they can go without food for a whole week.

123 The lioness' fur is sticky with the blood of its prey. Other lionesses of the pride help her clean herself off by licking her. The top of their tongue is covered with horned papilla, curved toward the back, which are useful for picking up food, licking blood and getting rid of parasites.

The cheetah hunts in daytime, thereby avoiding competition with other large felines. It is the only large predator that runs faster than its prey. Before hunting, it scrutinizes the horizon from a termitary or a tree stump. It often starts out in the open, picking out its prey among the animals which flee at its approach. Once it has made its choice, it leaps out at great speed. It must not begin running when it is more than 330 feet from its target or else the antelope or gazelle that it is stalking will have the time to rejoin the herd or escape, as it can run longer than the cheetah. After running 1300 feet at top speed the cheetah must rest as it exhausts itself in a few seconds. Its favourite victim is the Thomson gazelle. This feline is not a scavenger; it never returns to its prey and only eats what it has killed itself. The female always hunts alone unless it has young ones who are old enough to help her. The males sometimes get together to hunt bigger prey. The leopard, with its excellent camouflage and stealthy movements, can approach its prey without being seen and pounces by surprise. It cannot, however, run for a long time. Its keen sight, which is well-adapted for nighttime vision, enables it to hunt in the dark. Often, in order to avoid being robbed of its prey by other predators, the leopard carries it up a tree, even if it weighs more than 220 pounds. It then returns several days in a row to the same place to eat. The leopard's favourite prey are antelopes, gazelles, baboons and cattle. These large predators should not be in competition with one another as they have different preferences in terms of habitat and prey. Nonetheless, they are known to compete with and even hunt each other. Leopards have been spotted carrying cheetah carcasses into trees just as if they were impalas. As for cheetahs, they hunt and eat jackals. Lions kill and devour leopards and cheetahs; they also kill hyenas and jackals but do not eat them. The young of all predators are especially vulnerable to other species.

124-125 In periods of abundance, lion cubs do not lack for food, but in dividing up the killed prey, it's a case of the survival of the fittest and males are the first to eat their fill. Quarrels are quickly settled when food on the savannahs is plentiful, but in the dry season, even the smallest carcass is bitterly fought over. In these situations, hungry lion cubs cannot get to the food and often die of starvation.

126 top This cheetah had only just stopped to rest after the chase when a male baboon approached, attracted by the prey. Opening its jaws wide to reveal an impressive set of teeth, it intimidates the feline, which barely attempts to defend its property. The monkey charges straight at the cheetah, screaming, and forces it to abandon its meal.

126-127 This cheetah is sneaking up on a herd of impala. As soon as the feline pounces, the females run away, leaving the male behind to cover them while they escape.

127 top The cheetah is the fastest animal on land with a top speed of 110 km/h.

127 centre A mother cheetah carries a baby gazelle that is still alive to its six-month old young ones. They watch, not knowing what to do, then decide to chase it around playfully. This is their first lesson in how to hunt.

127 bottom The cheetah smothers its prey in order to kill it which is what it is doing here to a bohar reedbuck.

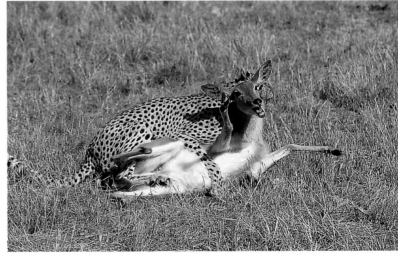

128 Leopards grab their prey by the throat and carry it into the trees. They can climb with up to 100 kilos (40 pounds) between their teeth. Their paws are tipped with very hard retractile claws that enable them to scale vertical tree trunks.

129 The leopard has an extraordinarily acute sense of hearing which helps them locate their prey. Its body is covered with extremely soft hairs which deaden the sound of its movements, thus enabling it to creep up on its prey undetected.

130-131 Crocodiles lie patiently in wait for their potential victims to approach the watering hole. They are caught, drowned, torn apart and stored in their "larder". These ferocious reptiles feast on the rows of gnus which cross the Mara river during migration. This is the only time that many of them eat.

132 Jackals are quite closely related to domestic dogs. They feed on prey that have been killed by large predators, and are thus in competition with vultures. They also hunt small mammals and eat reptiles, insects and fruit.

The prevalent image of the spotted hyena is that of a scavenger waiting for the big animals to finish their meal. Hans Kruuk has determined that the hyena is in fact a predator. Some hyena packs kill nearly 90% of the mammals which make up their diet. They always hunt in packs, often at night. They are fast and have great stamina and can maintain a speed of forty to fifty miles per hour for about 15 minutes. When one of them manages to catch a prey by its paw or flank, the others rush over to tear the tender flesh of the stomach until the victim collapses. Then the race for the spoils is on: they tear off huge chunks of meat, easily pulverizing the bones. This is usually the

132-133 Wild dogs attack the same types of prey as spotted hyenas but they hunt more efficiently, working together, whereas members of a hyena pack hunt more as individuals.

moment when the big predators come on the scene. Attracted by the yelping of the hyenas, they rush over and chase them away from their feast. Spotted hyenas can live in very different environments and survive in the most unfavourable conditions because of their powerful gastric juices and the high absorbency of their intestinal mucous, which enables them to digest bones and the excrement of other carnivores.
African wild dogs also hunt in packs. When they locate a herd, they sneak up on it with

their head down, neck extended, tail low and ears tucked in. They make no effort to hide, however, nor do they lie in wait or attempt to approach stealthily. The chase begins at about a 200 metre (650 foot) distance from the prey. The dogs head for the middle of the herd, choose one or a few of the weakest members and pursue them relentlessly, using a kind of relay technique. Though they cannot run as fast, they usually succeed because of their great stamina. They can rip a Thomson gazelle to shreds in seconds. Larger prey, like the gnu, are immobilised by several dogs' muzzles while others disembowel it and begin to eat it alive. They hunt early in the morning, late in the evening and sometimes when the moons is bright, and they only eat animals which they have killed themselves.

Their method of putting an animal to death may seem cruel but these canines do not possess the powerful weapons of felines; they do not have tapered claws nor the ability to close their fangs around the throat of their prey, as their necks measure only 60 centimetres (23 inches) in diameter. Though predators like lions, hyenas, wild dogs and jackals will occasionally scavenge dead animals, marabous, kites, and especially vultures are the principal cleaners of the savannah. Ungainly on the ground, these birds look splendid in flight: they soar high in the sky, carried by the hot air currents that rise from the savannah in the mid-morning. Vultures fly to great heights and their extraordinary view of the land below enables them to locate carcasses. These scavengers can safely eat decomposed, bacteria-ridden flesh because their extremely powerful digestive juices act as an antiseptic. Six main species of vultures coexist in East Africa. The African tawny vulture and the Ruppell vultures are by far the most numerous; in the Serengeti-Mara ecosystem, they represent 80% of all scavenger birds.

134 top The Seba python is the longest African snake, measuring between 3 and 5 metres long. A 4 metre python can swallow an animal weighing 25 kilos without any difficulty. When the python detects its prey, it hurls itself on top of it, seizes it between open jaws, and winds its body around it squeezing it to death. The jaw of a python can open to an angle of 130 degrees. It takes several days, even weeks, for a python to digest its prey and it can survive for two and a half years without eating.

134 bottom Big male baboons are particularly fond of new-born Thomson gazelles. They are always on the lookout for young prey though they are not exclusively carnivores. Their diet is varied and includes fruit, plants, roots, insects and birds.

134-135 This male
baboon, attracted by
the agonised cry of an
impala as it is being
killed by a leopard, rushes
over and steals the feline's
prey without any
opposition. In the dark,
however, felines gain the
upper hand over monkeys.
When night falls,
the wild cats rule.

Unable to cut into the skin, they use their long necks to probe animal's orifices. There are about 40,000 vultures in these two reserves and each year they consume over 12,000 tons of flesh.

This fact makes a nonsense of the phrase "he/she eats like a bird." Of course, in absolute terms, a small bird eats very little. But in relation to its size and weight, this represents a considerable quantity, disproportionate to anything we can imagine. Birds must get the most value possible out of their food and maintain a light weight. Flying and other incessant activities require much food; moreover, it is especially difficult to maintain a high and constant body temperature. This is particularly true for small species whose ration of absorbed food in relation to its body weight is astonishing. It has been calculated that to eat proportionally as much as a humming-bird, the smallest of all bird, a man must consume on a daily basis 65 kilograms (140 pounds) of bread and 80 kilograms (400 pounds) of potatoes. Birds have an infinite variety of diets. There are: herbivores, like geese; frugivores, like parrots and the palmist vulture; grain-eaters; nectar-drinkers; and connoisseurs of molluscs, fish and insects. Some feed off land vertebrates, including other birds, while others filter mud, like flamingos. There are even birds that use tools and amass provisions, like shrikes. In exchange for the food that plants provide, birds render reciprocal services. Nectar-loving birds, for example, distribute pollen from the flowers they feed off which they carry on their heads and beaks. They deposit it on the pistils of subsequent flowers they visit, thus fertilising them. Other birds help by dispersing seeds.

136-137 Like other vultures, the oricou is a scavenger, but it is also a predator which likes to hunt flamingos and young gazelles. Around carcasses, they tend to stay on the periphery and chase other vultures away by charging at them with their head lowered, neck extended and their back feathers ruffled. Vultures are rarely seen early in the morning as most of them wait for the sun to warm up the air, creating ascending thermal currents that they use to fly.

138 The secretary-bird is a special bird of prey which looks very much like a wader. It has long legs and neck, is land-based and eats snakes.

138-139 The Egyptian vulture, with its long and robust beak, can pick up stones and use them as tools, a unique characteristic among birds. By throwing stones at an ostrich egg, it breaks the shell, which is several millimetres thick.

139 top Cranes live in couples or small groups in the open plains or swamps. They eat seeds, insects and frogs. Some tribes believe that these birds bring the rain.

*142-143 The lioness'
silhouette against the sky
is lit up by the sunset.
Night has fallen, the time
when lions rule. But these
gazelles have nothing to
fear because a lion out
in the open has no chance
of outrunning a healthy,
alert antelope.*

*144 This litter of
cheetahs on the plains
outside the Masai-Mara
reserve is fortunate
because few hyenas and
lions live here.
The Masai killed a great
many by poisoning
carrion to protect their*

*herds from large
predators. As cheetahs
only eat what they kill
themselves, they have
nothing to fear.
Consequently the
concentration of young
cheetahs in this area is
relatively high.*

*140-141 The fisher
eagle's cry is instantly
recognisable. It sits on
its perch overlooking the
water looking for its prey.
When it spies a fish,
it swoops down on it, grabs
it with its claws and goes*

*back to its perch. It spends
little time fishing,
however, preferring to
engage in pirate activity,
attacking and robbing
pelicans, herons,
kingfishers and even other
birds of their prey.*